MIX
Papier aus verantwortungsvollen Quellen
Paper from responsible sources
FSC® C105338

Dr. Perumalla Janaki Ramulu
A. Lavanya

Design and Fabrication of Equal Channel Angular Extrusion Process Analysis for Non-Ferrous Materials

Anchor Academic
Publishing

Janaki Ramulu, Perumalla, Lavanya, A.: Design and Fabrication of Equal Channel
Angular Extrusion Process Analysis for Non-Ferrous Materials, Hamburg, Anchor
Academic Publishing 2017

Buch-ISBN: 978-3-96067-106-0
PDF-eBook-ISBN: 978-3-96067-606-5
Druck/Herstellung: Anchor Academic Publishing, Hamburg, 2017

Bibliografische Information der Deutschen Nationalbibliothek:
Die Deutsche Nationalbibliothek verzeichnet diese Publikation in der Deutschen
Nationalbibliografie; detaillierte bibliografische Daten sind im Internet über
http://dnb.d-nb.de abrufbar.

Bibliographical Information of the German National Library:
The German National Library lists this publication in the German National Bibliography.
Detailed bibliographic data can be found at: http://dnb.d-nb.de

Dr. Perumalla Janaki Ramulu and A. Lavanya
M.Tech (NIFFT), (PhD) (IITG)
MIE, IACSIT (MEG), IAENG, SCIEI, SMFRA, IDDRG, MASME, UAMAE, WASET, SAI Mech,
ESME, ISAET
Associate Professor in the Programme of MDME, SoMCME,
Adama Science and Technology University,
Adama, Post Box: 1888, Ethiopia

All rights reserved. This publication may not be reproduced, stored in a retrieval system
or transmitted, in any form or by any means, electronic, mechanical, photocopying,
recording or otherwise, without the prior permission of the publishers.

Das Werk einschließlich aller seiner Teile ist urheberrechtlich geschützt. Jede Verwertung
außerhalb der Grenzen des Urheberrechtsgesetzes ist ohne Zustimmung des Verlages
unzulässig und strafbar. Dies gilt insbesondere für Vervielfältigungen, Übersetzungen,
Mikroverfilmungen und die Einspeicherung und Bearbeitung in elektronischen Systemen.

Die Wiedergabe von Gebrauchsnamen, Handelsnamen, Warenbezeichnungen usw. in
diesem Werk berechtigt auch ohne besondere Kennzeichnung nicht zu der Annahme,
dass solche Namen im Sinne der Warenzeichen- und Markenschutz-Gesetzgebung als frei
zu betrachten wären und daher von jedermann benutzt werden dürften.

Die Informationen in diesem Werk wurden mit Sorgfalt erarbeitet. Dennoch können
Fehler nicht vollständig ausgeschlossen werden und die Diplomica Verlag GmbH, die
Autoren oder Übersetzer übernehmen keine juristische Verantwortung oder irgendeine
Haftung für evtl. verbliebene fehlerhafte Angaben und deren Folgen.

Alle Rechte vorbehalten

© Anchor Academic Publishing, Imprint der Diplomica Verlag GmbH
Hermannstal 119k, 22119 Hamburg
http://www.diplomica-verlag.de, Hamburg 2017
Printed in Germany

ABSTRACT

"Equal Channel Angular Extrusion (ECAE)" is a significant method in industrial forming applications, which is the most important method for production of ultrafine grained bulk samples, high plastic strains are introduced into the bulk material without any change in cross section. Equal channel angular extrusion has different die channel angles from which an optimum die channel angle should be identified. So that competent mechanical properties will be obtained from the extruder. This work is focused on the plastic deformation behavior of Al alloys by developing ECAE process and also studied the finite element analysis. For the simulation, the whole ECAE setup was modeled using CATIA and converted into STL file format. The STL files of ECAE imported into DEFORM-3D for simulations. The experiments are performed by fabricating the ECAE tools such as die, punch and billet. A series of experiments were carried out for the die angles of 115°, 125°and 135° and outer corner angle of 6°, billet diameter 9mm and height 70mm was used. A detailed analysis of the strains introduced by ECAP in a single passage through the die is noted. The experiments were conducted by attaching the ECAE tools to the Universal Testing Machine on aluminum alloy. From the experiment and simulation results, load, displacement, and punch force are evaluated and compared with each other.

TABLE OF CONTENTS

ABSTRACT i

TABLE OF CONTENTS ii

LIST OF FIGURES iv

LIST OF TABLES v

CHAPTER 1: INTRODUCTION 1

    1.1 History of Equal Channel Angular Extrusion Process 1

    1.2 Aim and Scope of the Study 2

    1.3 Equipment Used 3

        1.3.1 Hydraulic Press 3

        1.3.2 Mechanical Press 3

    1.4 Tasks Involved 3

    1.5 Organization of the Study 3

CHAPTER 2: LITERATURE SURVEY 5

    2.1 Experimental Studies on Equal Channel Angular Extrusion Process 5

    2.2 Numerical Analysis of Equal Channel Angular Extrusion Process 11

    2.3 Finite Element Analysis of Equal Channel Angular Extrusion Process 13

    2.4 Other Studies on Equal Channel Angular Extrusion Process 18

CHAPTER 3: METHODOLOGY 19

    3.1 Experiment Setup 19

    3.2 Equipment 20

        3.2.1 Machine Frame or Loading Unit 20

        3.2.2 Hydraulic System Unit 21

        3.2.3 Electronic Control Unit 21

    3.3 Die Design Details 21

    3.4 Die Design Consideration 22

    3.5 Different Parts of the Setup 22

3.6 Equal Channel Angular Extrusion Process Parameters 24

    3.6.1 Channel Intersection Angle (Φ) 24

    3.6.2 Outer Corner Angle (Ψ_0) 24

    3.6.3 Inner Corner Angle (Ψi) 24

    3.6.4 Inner Corner Radius (R_i) 24

    3.6.5 Outer Corner Radius (R_o) 24

    3.6.6 Temperature of Billet and Die 25

    3.6.7 Friction 25

    3.6.8 Force 25

    3.6.9 Number of Passes 25

    3.6.10 Deformation Route 25

    3.7 Modeling of Equal Channel Angular Extrusion Setup 26

    3.8 Raw Material Used For Experiment 27

    3.9 Experimental Procedure For ECAE 28

    3.10 Finite Element Analysis 29

    3.11 Simulation Methodology 30

CHAPTER 4: RESULTS AND DISCUSSION 31

    4.1 Experimental Investigation 31

    4.2 Stress Evaluation at Different Steps 31

    4.3 Punch Force Evaluation 34

CHAPTER 5: CONCLUSIONS AND FUTURE SCOPE 38

REFERENCES 39

LIST OF FIGURES

3.1	Schematic representation of ECAE setup	19
3.2	Universal Testing Machine fixtures	20
3.3	UTM Hydraulic system unit	21
3.4	Punch and different channel angles of die	23
3.5	Modeling of punch	26
3.6	Modeling of die	26
3.7	Assembly of ECAE setup	27
3.8	ECAE setup on Universal Testing Machine	28
4.1	Before and after ECAP processing of the Al samples	31
4.2	Stress formations at different progressions for die channel angle of 105°	32
4.3	Stress formations at different progressions for die channel angle of 115°	32
4.4	Stress formations at different progressions for die channel angle of 125°	33
4.5	Stress formations at different progressions for die channel angle of 135°	34
4.6	Comparison of extrusion load with displacement with die channel angle of 105°	34
4.7	Comparison of extrusion load with displacement with die channel angle of 115°	35
4.8	Comparison of extrusion load with displacement with die channel angle of 125°	35
4.9	Comparison of extrusion load with displacement with die channel angle of 135°	36

LIST OF TABLES

3.1	Tool dimensions and processing parameters	22
3.2	Material properties of the aluminum	28
4.1	Die channel angles and their obtained results	37

CHAPTER 1
INTRODUCTION

Equal channel angular extrusion (ECAE) is a technique for producing ultra fine grain structures in submicron level by introducing a large amount of shear strain into the materials without changing the billet shape or dimensions. Ultra-fine grained (UFG) materials, having grain sizes in the sub-micrometer range, have always been the focus of interest and extensive research in materials science. Conventional heavy deformation techniques such as drawing and cold rolling are also accompanied with significant refinement in the microstructure. The most important reasons for this interest and corresponding research activities are the unusual mechanical and physical properties that UFG materials possess, when compared to traditionally produce coarse-grained materials. This interest has grown significantly in the recent years with the development of several plastic deformation (SPD) methods. Which provides the opportunity to produce bulk UFG materials without difficulties encountered during the previous fabrication methods. Equal channel angular extrusion (ECAE) is one of the most promising methods in SPD methods, since the workpiece cross section does not change during the process. The details of ECAE are discussed in the following sections.

1.1 History of Equal Channel Angular Extrusion Process

The equal channel angular extrusion (ECAE) was developed by Segal in 1981. Now the process is extensively used to achieve ultra-fine grained structures in bulk form, which gives rise to attain the outstanding properties such as high strength, toughness and hardness at ambient temperatures, exhibition of superplastic behavior at high strain rates at elevated temperatures.

In ECAE process, two channels of the same cross section intersect at a certain angle to form the die. The billet either round or square cross section, is pushed from the top into this die by means of a flat faced ram and is subjected to simple shear without any change in cross sectional dimensions. By repeated pressings, the work piece can attain very large strains which are almost impossible to obtain via conventional manufacturing methods.

Equal channel angular extrusion processed parts can be used in industrial applications, aerospace and automotive industries, the products include fasteners like screws, screw rivets used in the assembly of aluminum components for aircraft and other structures, elements for aircraft fuselages (stringers, skin plates, etc) etc.

Equal channel angular extrusion is also strain hardening process. Strain hardening (also called work hardening or cold working) is the process of an increasing stress; stress is required to produce plastic deformation and the metal apparently becomes stronger and more difficult to deform. When a metal is plastically deformed, dislocations move and additional dislocations are generated. The more dislocations within a material, they will interact more and become pinned and tangled. This will result in a decrease in the mobility of the dislocations and a strengthening of the material. This type of strengthening is commonly called cold-working. It is called cold-working because the plastic deformation must occur at a temperature low enough that atoms cannot rearrange themselves. When a metal is worked at higher temperatures (hot-working) the dislocations can rearrange and little strengthening is achieved.

Equal channel angular extrusion is extensively used in the bulk metal forming under compressive conditions, without altering the shape or dimensions of the billet. The formability of bulk material during ECAE process depends on the parameters like lubrication, punch force, die inner corner angle and outer corner angle, die-punch clearance, in addition to mechanical properties and shape of the billet and parts geometry. The proper selection of these variables helps to reduce the defects in the ECAE process. It also helps to implement the optimum results and improves the mechanical properties of the materials used in the process.

When it comes to material processing for industrial and research applications, the light metals will have a great amount of interest. The property of lightness translates directly to the material property enhancement for many products since by far the greatest weight reduction is achieved by a decrease in density. The term "light metals" has traditionally been given to aluminum because they are frequently used to reduce the weight of components and structures. This is quite expected since specific strength becomes one of the most important desired properties and is easier to be achieved due to the low densities of such materials.

1.2 Aim and Scope of the Study

In this study, the effects of equal channel angular extrusion process parameters investigated by both experimentally and numerically. The study is initiated by modeling ECAE tools. ECAE process parameters namely different die channel angles, outer corner angle; numbers of passes, punch force are studied. The experiments are carried out with the help of universal testing machine by fixing the ECAE tools setup. The simulations are performed in finite element code by importing the tool into FE code. The dimensions of the ECAE tool, the size of the billet, properties of material, the varied process parameters and power law criteria considered for simulations are given in Chapter 3 elaborately. From the simulations die channel angles, punch force, load, displacement are evaluated for all conditions. From the overall analysis of the results, optimizations of the ECAE process parameters for different die channel angles are suggested and produce maximum property improvement.

1.3 Equipment Used

The equipment on the basis of the type of force used to drive the punch is classified as follows:

1.3.1 Hydraulic Press

In this type of equipment the force required to derive the ram (punch) is supplied by hydraulic means to deform the metal plastically

1.3.2 Mechanical Press

In this type of equipment the force required to derive the ram (punch) is supplied by mechanical means to deform the metal plastically.

1.4 Tasks Involved

- Design, Modeling and Fabrication of whole laboratory level ECAE process setup.
- Perform the ECAE process simulations using DEFORM 3D a finite element code.
- Conduct the ECAE process experimentation on universal testing machine.
- Investigate the effect of different die angles on ECAE process.
- Obtain the forming behavior of aluminum 1100 material at different die angles 115°, 125° and 135°.
- Compare and validate Experimental results and simulation results.
- Optimize die channel angles to obtain ultrafine grain structure and improving mechanical properties of the aluminum alloy.

1.5 Organization of the Study

The study is divided into five chapters. In the first chapter, general information about the equal channel angular extrusion process and outline of the present study is given. The next chapter is the literature review, in which previous studies on equal channel angular extrusion process, experimental results, numerical simulations and the analysis are summarized. The finite element analysis and experimental applications of the ECAE process are overviewed. In the third chapter the factors influencing the equal channel angular extrusion process, the effect of various process parameters of the ECAE process, the use of finite element analysis for numerical simulations and experimental investigations are explained. Fourth chapter focused on representation of the results obtained by varying the process parameters. The fifth chapter deals with the conclusion of the ECAE of aluminum with different die angles.

CHAPTER 2
LITERATURE SURVEY

In this chapter, a few research papers have been indicated related to present work objective. The investigations provided the vast amount of information about the Equal channel angular extrusion process. From the various studies, they suggested many approaches on ECAP. In this section, a survey on experimental works, numerical simulations and finite element analysis of equal channel angular extrusion process has carried out.

2.1 Experimental Studies on Equal Channel Angular Extrusion Process

From a decade, there were many studies carried out on experimental observations on ECAP and various parameters involved in the process. For example, Segal *et al* (1999) studied about the plastic flow during the one step and evolution of shear planes during multi pass extrusion. In this work, pure aluminum was processed through equal channel angular extrusion with low friction and tool angle of 90°. The results found that the equivalent strains cannot be unique between ECAE and ordinary forming operations. Each pass of ECAE process was developed thin shear bands along the channels. Segal (2003) investigated about the equal channel angular extrusion by slip line method. In this analysis, it is noted that contact friction was important effect on stress-strain state during ECAE. When the friction increases from zero to maximum, the strain distribution remains uniform and effective strains reduced. Wei *et al* (2003) studied about the microstructure and mechanical properties of tantalum after equal channel angular extrusion process with four passes. After completing the ECAE processed material was tested for under quasi-static and dynamic loading conditions. The result found that, the tantalum shows perfectly plastic behavior under quasi-static loading without strain hardening. In dynamic loading condition, slight softening due to adiabatic heating. Markushev *et al* (2004) studied about the structure and mechanical properties of commercial Al-Mg 1560 alloy after equal-channel angular extrusion. The sub microcrystalline deformation structure has obtained by ECAE and it was transformed into sub-micron and microcrystalline grain structure upon annealing. The nature of recrystallization and phase transformation processes were discussed. Mechanical properties such as strength, hardness, ductility and crack resistance of the alloy were demonstrated. The result showed that ductility and crack resistance were decreased with transformation of the SMC structure. Wang *et al* (2004) studied about the effect of deformation temperature on the microstructure developed in commercial purity aluminum by equal channel angular extrusion process. The authors have demonstrated the effect of temperature on the microstructures was investigated after ECAE. They characterized the micro structural parameters, including grain size, shape and boundary misorientation. It is noted as sub grain size was increased due to increase the temperatures from 289-523K.

Saravanan *et al* (2006) studied about the equal channel angular pressing of pure aluminum to produce ultra-fine grain structures in submicron level. The ECAP technique was attempted to 99.5% pure aluminum. The aluminum alloy was characterized by optical metallography, atomic force microscopy and hardness measurement. The result found that, the number of passes important, to achieve a homogeneous microstructure in pure Al. The ultimate equilibrium grain size was obtained through ECAE process. Fang *et al* (2006) studied about the effect of equal channel angular pressing on tensile properties and fracture modes of casting Al-Cu alloys. In this work the grains of the two alloys refined to submicron level after four passes of ECAP. The result found that, the tensile fracture strength and static toughness was increased with ECAP pass. The failure modes of two Al-Cu alloys were exhibit different features, such as necking degree decreases and shear feature becomes more with increasing of ECAP. Gazder *et al* (2006) studied about the progressive texture evolution during equal channel angular extrusion process. The progressive texture evolution was seen during second pass of equal channel angular extrusion of copper. In this work, the uniform texture predictions were achieved by visco-plastic self-consistent. This process was using finite element predicted deformation. The result found from above work, the greatest texture changes were occurred in the billet. It was found in a narrow region between the ends of the entry channel and die intersection plane. Nagashekar *et al* (2006) studied the ECAP on tubular aluminum alloy specimens. The experimental test was conducted to measure the pressure and the mechanical properties of the extruded material. In this work, low extrusion pressures during ECAE of tubular specimens are due to the movement of the mandrel along with the specimen. During this work the tensile strength, yield strength, and hardness were improved and elongation to failure was decreased. Finally this technique was improved properties of the tubular specimen.

Yeung *et al* (2007) studied about the equal channel angular extrusion of high purity gold. In this work, the high purity gold billet of 10mm diameter and it was processed up to 9 passes. The result found that, the hardness of the gold sample was increased from 30HV to 82 HV. The electron backscatter diffraction was showed an average grain size 140nm in the extruded material. The high volume of high angled grain boundaries was determined in the extruded gold solid. Suzuki *et al* (2008) studied about the equal channel angular extrusion process of lotus-type porous copper. This work was investigated using ECAE die with different die angles. The copper rod was densified by the uniaxial compression in the entry of the channel. The pores were thinned elongated by shearing at the corner of the die. The result found for this analysis was, due to the shear flow stress reducing the normal stress. The pore morphology was controlled by the extrusion process. The Vickers hardness increased through the ECAE process and improved the mechanical properties of porous metals. Paydar *et al* (2008) studied about the consolidation of Al particles through forward extrusion-equal channel angular pressing. In this work two processes were conducted in a single die, such as forward extrusion and equal channel angular pressing. The forward extrusion-ECAP was superior to conventional forward extrusion. This work was to achieve full density and good bonding in entire volume of the consolidated particles. The result of this FE-ECAP process, the bulk material has high strength and excellent ductility. The advantage of this method was possibility of performing two processes in a single die and producing a long bar samples with full density. Gao *et al* (2008) studied about the microstructure and dry sliding wear behavior of Cu-10%Al-4%Fe alloy produced by equal channel angular extrusion process. The aluminum bronze alloy was processed by ECAE at higher temperature. Authors were investigated the effect of microstructure, mechanical and tribological properties of the alloy was investigated. The result found, the grains were refined and grain size was decreased after ECAE. The friction coefficient of the specimen was decreased with decreasing of grain size. The grain refinement can improved wear resistance and the load bearing capacity of the alloy during dry sliding. Hardness and strength of the alloy was increased due to increase the ECAE passes.

Daly *et al* (2009) studied about the effect of annealing on the microstructure and properties used equal channel angular extrusion. In this work, 99.99 wt. % of the pure copper was processed at room temperature. The ECAE technique was processed in the material, without changing the sample dimensions through ECAE process. In this work, the micro structure and micro hardness were studied. The result found that, increased annealing temperature, the microstructure has becomes more homogeneous while the micro hardness become decreased. Atef Rebhi *et al* (2009) studied about the characterization of aluminum processed by equal channel angular extrusion: effect of processing route. In this work, an aluminum alloy was processed by ECAE via routes B_c and C up to four passes. This process was characterized the evolution of the microstructure and changes of mechanical properties. The result of this work was the tensile test showed; the routes B_c and C lead to higher flow stress and lower strain-hardening coefficient after ECAE. The dynamic recovery was more effective in route C than route B_c due to high density of dislocations. Chinh *et al* (2010) studied about the processing of Age-hardenable alloys by an equal channel angular pressing at room temperature. In this work, the microstructure, strength, and ductility of age hardenable Al Zn Mg alloys was investigated. The investigation was reported, a critical strategy for the processing of age-hardenable alloys by conducting the quenching. The result of this work was improving the strength, reduced the possibility of the formation of microcracks during several plastic deformations by using ECAP. Further, the application of ECAP process may increase not only the strength but also the ductility of the age-hardenable samples after a certain number of ECAP process. Biswas *et al* (2010) studied about the room-temperature equal channel angular extrusion of pure magnesium. This work was conducted at room temperature, to produce ultrafine grain size through ECAE. The method of this analysis was, to achieve required grain sizes had proposed. It was to obtain suitable initial orientation with high Schimed factor for basal slip. To take advantage of low stacking fault energy of basal and high stacking fault energies of prismatic plane. The temperature deformation was lower in steps, leading to continual refinement of grains. The result of this work was the hardness of the material increased with decrease in the grain size according to hall-pitch relationship.

Vasile Danut *et al* (2010) studied about the mechanical behavior comparison between unprocessed and ECAP processed 6063-T835 aluminum alloy. The ECAE method was used to modify the microstructure and producing ultra-fine grained metals and nano materials. This work was to investigate the mechanical behavior of the ECAP processed material by uniaxial tensile test. This work was the ultimate tensile strength, yield strength, and strength to fracture was observed after ECAP. The result of the work was found, the ECAP process was to increase the mechanical properties. Dong-Hwan *et al* (2010) studied about the mechanical behavior and micro structural evolution of commercially pure titanium. The T-type ECAP apparatus was developed to improved the efficiency of common ECAP and the finished products. CP-Ti billet was much stronger; it was produced by using the ECAP process. The result of this work was decreased the grain size occurring through the multi pass ECAP caused by SPD. In this process spatial variations of microstructure together with relevant mechanical properties were predictable. The material status in the plastic forming process coupled with the processing condition and particularly focused to determine the properties of finished products. Orlov *et al* (2011) studied about the mechanical properties of magnesium alloy ZK60 by combines conventional extrusion and equal channel angular pressing. In this work, only single pass process of ECAE was done. The result of this work was microstructure and texture of the SPD-processed bars showed that the excellent combination of strength and ductility. The excellent balance of enhanced mechanical properties was achieved with ZK 60 alloy. Eslami *et al* (2011) investigated about the diffusion bonding of aluminum to copper using equal channel angular extrusion process. Diffusion bonding process was new method for production of bimetallic rods; utilizing ECAE process had been introduced. In previous analysis no attempt had been made to assess the effect of different temperatures and holding times in order to achieve the diffusion bond between the mating surfaces. In this work the copper sheeted aluminum rods were extruded at constant pressure, different temperatures and holding times to produce a diffusion bond between the copper sheeted aluminum cores. The result of this analysis was the bonding temperature of 200^0 c and holding time of 60-80 min yielded the highest shear strength value was found. At constant temperature the joint strength was increased with increasing the holding time.

Kazeem *et al* (2012) studied about the equal channel angular pressing technique for the formation of ultra-fine grained structures. In metal forming process EACP is one of the technique, in which an ultra large plastic strain was imposed on bulk material in order to make an ultra-fine grains. This forming technique was to extrude the material by use of specially designed channels without changing the die geometry by imposing the several plastic deformations. The samples were subjected to hardness test; the results showed improve the mechanical behavior of the ultra-fine grains in copper alloy. Mohan reddy *et al* (2013) studied about the improving the mechanical properties of Al 7075 alloy by equal channel angular extrusion process. In ECAE analysis the microstructure and tensile properties were investigated. The process was attempted at room temperature. The result of this analysis was sub grains increased and the width of boundaries decreased while the sub grain size remains approximately constant as the number of passes increases. The Tensile test was showed, to increase the strength from one pass to another pass of the material. Hardness also increased by increasing the number of passes due to the fine grain structure. Li (2013) studied about the application of crystal plasticity modeling in equal channel angular extrusion process. In the equal channel angular extrusion process the crystal plasticity modeling and face-centered cubic metals were highlighted. The equal channel angular extrusion process was demonstrated, the crystal plasticity models were used in exploring the crystallographic nature of grain deformation. The result found from the above work, the simulations was dependency of grain refinement efficiency on processing route. This can be capture the orientation stability and texture evolution under various processing conditions. Fan *et al*

(2013) studied about the microstructure and mechanical properties of T91 steel processed by ECAP at room temperature. The T91 steel was annealed at $500°C$ for 2 hours subsequently. The result of this analysis was the number of passes increasing; the grain size was gradually refined. The minimum average grain size was 200 nm, when the sample was extruded after six passes at room temperature. The micro hardness and tensile strength were increased with the number of extrusion passes increased. The corresponding elongation was decreased. In the annealing process, the average grain size was increased. During the ECAP process, the samples were improved the ductility and little decrease in tensile strength. Hao *et al* (2013) studied about the microstructure and properties of ultrafine-grained tungsten produced by equal-channel angular extrusion. The ultra-fined grain tungsten was fabricated by equal channel angular pressing at different temperatures. The ECAE sample was characterized by electron microscopy and Vickers micro hardness measuring instrument. The result of this analysis was the average grain size of width was decreased with the increasing number of passes. In this process intergranular fracture mode dominates the fracture failure. The micro hardness of tungsten was increased and the thermal conductivity was affected only a little by ECAP process.

Bouaksa *et al* (2014) studied about the molecular chain orientation in polycarbonate during equal channel angular extrusion. The test was conducted on glassy amorphous polymer. The experimental test was characterized the microstructure of the ECAE deformed material. The deformation character of the polycarbonate specimen in the ECAE process was simulated using visco hyper elastic-viscoplastic constitutive model, which was implemented in a finite element code. This analysis was to verify the simulated load-displacement curve compared with experimental data. The result was obtained; the simulations were predicting a high degree of anisotropy and heterogeneity in orientation of polymer molecular chains.

Tong *et al* (2014) investigated the microstructure and Martensitic transformation of an ultrafine-grained TiNiNb shape memory alloy. The ultrafine-grained $Ti_{44}Ni_{47}Nb_9$ shape memory alloy processed by ECAP at $450°c$ for eight passes was investigated. The ECAP processed sample was characterized by an inhomogeneous and refined microstructure. In this shape memory alloy the β-Nb phase-rich region, the grains of matrix were elongated with high density dislocations. In β-Nb phase-free region, the microstructure was partial recovery and characterized by near-equiaxed grains. The heterogeneous microstructure was attributed to β-Nb phase region. Martensitic transformation behavior of the ECAP processed sample was characterized by a single transformation. Lower transformation temperature was compared to the initial sample. The thermal cycling stability of transformation and the mechanical properties were improved due to a strengthening effect resulting from refined grain size and high dislocation density. Thiyagarajan *et al* (2014) studied about the enhancement of mechanical properties of AA6351 using equal channel angular extrusion. The die was consists two channels of equal cross section intersecting at an angle of $110°C$. The work piece was placed in one channel and extruded into the other channel using a punch. The ECAE process was conducted in room temperature. The result of this analysis was the tensile strength of the alloy also increased with increase in number of passes, but there is some reduction in ductility due to the hardening of the alloy. The micro hardness of the alloy was increased with increased in number of passes and attains maximum value of 105 VHN after two passes the formation of fine grains and high density dislocations.

2.2 Numerical Analysis of Equal Channel Angular Extrusion Process

There were a few numerical studies on the ECAE process discussed. For example, Cui *et al* (1998) studied about the three-dimensional simulation of flow pattern in equal channel angular extrusion process. They developed the model to extrude the material without changing the cross-sectional area. In this process, by changing the direction of the billet, texture and microstructures of the material were developed in the material. The ECAE process was according to orientation of four fundamental passes were defined and used for different purposes. It was important to understand the flow pattern after each pass. In this analysis numerical methodology was used to define the flow patterns. Accordingly, the numerical simulation visualization program was developed to simulate the flow pattern for any route and number of passes.

Beyerlein *et al* (2004) studied about the analytical modeling of material flow in equal channel angular extrusion process. In this work, the analytical forms of deformation and velocity gradients associated with the single pass of ECAE. The three stages of plastic deformation was considered such as ideal simple shear, plastic deformation zone in the shape of central fan deformation and final deformation of initial cube and sphere. The benefits of the analysis was, the initial deformation was well constrained within the die, and details of texture or dislocation configurations can be studied using poly crystal models without having to couple them to FEM calculations. Aour *et al* (2006) studied about the numerical investigation on equal channel angular extrusion process of polymer. Various geometrical parameters such as channel angle, outer corner angle, inner radius, material parameters, plastic deformation and strain homogeneity of polymer were studied in the numerical analysis. The result found that, the strain was decreased from the inner part to outer part. The pressing load of this analysis was increasing the accumulating of plastic strain up to stability of the material flow. The friction between the surface of the material and die was modeled with Coulomb friction law and is assumed to be uniform everywhere in the die. Zairi *et al* (2006) was studied about the numerical modeling of elastic-viscoplastic equal channel angular extrusion process of a polymer. In this process the plastic response of a polymer at room temperature had investigated by numerical simulation. The main objective of this work was basic understanding of the plastic flow in the polymer during one ECAE pass. The distribution of strain, strain rates and stresses, the deformation behavior of sample and the load-displacement curves was analyzed in numerical modeling of elastic-viscoplastic model. The strain homogeneity was studied for different tool angles by considering the influence of realistic material parameters. The result found from above analysis was, strain distribution was strongly dependent on the material behavior.

Eivani *et al* (2007) studied about the effective strain based on shear and principal strains in equal channel angular extrusion with outer curved corner. In this study, two relationships were derived to calculate the effective strain based on shear and principal strains. First solution was evaluating the effective strain in multi-stage ECAE dies consisting of two or more sub-dies. The first solution was extended, to calculate the effective strains in ECAE dies with outer curved corner. The result of these two solutions between experimental and theoretical results increased as the outer curved corner angle increased. Eivani *et al* (2007) studied about the effect of dead metal zone formation on strain and extrusion force during the equal channel angular extrusion. In this analysis investigate the deformation of material during equal channel angular extrusion process. This analysis was used an upper bound model. In this analysis friction between the sample and die walls on the geometry of dead metal zone, effect of die angle and total strain was derived. The result found from the above analysis was, the strain decreased with the increasing of die angle and friction coefficient. The deformation zone was develops with the increasing the friction coefficient and decreasing the die angle. Bajor *et al* (2014) studied about the numerical analysis of AZ61 magnesium alloy extrusion process by modified equal channel angular extrusion method. It is a hot deformation method. The extrusion process was using the die with modified angular channel containing horizontal contracting zone in the material exit direction. This modified ECAE final product was round rod that could be used as a material charge to further plastic deformation process. The main aim of this process was to determine the most favorable parameters of the extrusion that allow obtaining the products with good mechanical properties. The result found from this analysis was, to increase the deformation rate the strain hardening of examined alloy grows, where the temperature growth causes the lower level of yield stress in compression test.

2.3 Finite Element Analysis of Equal Channel Angular Extrusion Process

Kim *et al* (2000) studied about the die corner gap formation in equal channel angular extrusion. The finite element analysis DEFORM 2D code was used to investigate the corner gap formation between die and workpiece during plain strain ECAP process. The comparison of the deformation and die corner gap formation behavior between strain hardening material and quasi perfect plastic material was made. The result found, the larger gap formation was formed in the material with higher strain hardening rate because the softer outside part of the workpiece in the deforming region flows faster in the strain hardening material. The corner gap formation reduced the strain in the outside region of the workpiece, increased the strain in the inside region and decreased the average strain. The strain distribution of the workpiece with larger die corner gap becomes more inhomogeneous. Li *et al* (2004) analyzed the formation of the plastic deformation zone (PDZ) and evolution of the working load along with the ram displacement in a single pass of equal channel angular extrusion (ECAE) with an intersection angle of 90°. This study explored systematically coupled with the effects of material response, outer corner angle ($\Psi = 0°$, $45°$, or $90°$), and friction on ECAE deformation. Effective strain calculations are compared with various analytical models and it is directly an account for the PDZ tends to perform better. Nagsekhar *et al* (2004) studied about the optimal tool angle for equal channel angular extrusion of strain hardening materials by finite element method. This deformation technique was imported the large amount of plastic strain in to the bulk material through the application of uniform simple shear. This process can be carried out by finite element code ABAQUS/explicit. This analysis was used to analyze the deformation behavior of extruded materials and strain homogeneity was studied for different tool angles. By considered, the influence of realistic parameters like, strain hardening behavior and frictional contact between the die and sample. The result found, the optimal strain homogeneity in the sample with lower dead zone deformation, without involving any effects, can be achieved with channel angle of 90^0 and outer corner angle of $10°$.

Nagasekhar *et al* (2005) investigated about the stress and strain histories in equal channel angular extrusion/pressing. In this analysis several plastic strains were induced in the material through multiple equal channel extrusion. It can produce bulk ultrafine grains and it was suitable for structural applications. In this study to design an optimized ECAE die, the stress-strain histories, and punch force requirements are very important. This work was carried out by the finite element analysis code was ABAQUS/Explicit for a range of different channel angles. The result found, the peak punch force decreased gradually with the increase in channel angle. Son *et al* (2006) studied about the finite element investigations of friction condition in equal channel angular extrusion. This study was mainly focused on the investigation of contact phenomenon at the interface between dies and the work piece. In this, the material flow and forming load were main requirement. In this used methodology was numerical simulation of an ECAE with a CP-TiGr-1 cylindrical specimen were carried out by applying the mixed finite element formulation with tetrahedral elements under non isothermal condition. Compression test was used to determination of material data. The result found from the analysis, the forming loads varied very sensitively depending on the friction conditions. Luri *et al* (2006) studied about the new configuration for equal channel angular extrusion dies. This process was used to impart severe plastic deformations to processed the material and improving the properties of the materials and reducing the grain size. In the present study discuss the configuration of new die proposed for ECAE process. This new configuration of die has more advantage compare to conventional die. The new die was obtaining higher plastic strain in each passage than conventional die. The optimization of die geometry was important using finite element methodology. This analysis was determined how the variations are effected on die geometry. The result found, both finite element method and analytical methods will allow us to affirm that by using this new die configuration it is possible to achieve higher deformation values per ECAE passage.

Zairi *et al* (2006) studied about the influence of the initial yield strain magnitude on the material flow in equal channel angular extrusion process. In this analysis using methodology was finite element method. This process was to simulate the material nonlinearity characteristics such as strain hardening, plastic strain ratio, and strain softening and die geometry on the deformation behavior of a sample were simulated. The result found from this analysis was, the plastic strain heterogeneity and sample bending were increases with initial yield strain. The maximum processing load was decreased. Nagasekar *et al* (2007) studied about the deformation behavior and strain homogeneity in equal channel angular extrusion. In this process, finite element code ABAQUS was used. In ECAE process several plastic deformation techniques were widely used for fabrication of bulk nano structured materials, powder materials and tubular materials. The deformed materials microstructure and mechanical properties were strongly depended on the amount of strain induced and strain homogeneity. Strain homogeneity and deformation behavior was most important to design an ECEA die. The result found, the effective strain variation across the width at the centre of the sample showed that strain homogeneity was greater. Cheng *et al* (2008) studied about the investigation of equal channel angular extrusion process of billet with internal defects. In this mainly focused on the shear plastic deformation behavior of a material during ECAE process. This process was governed primarily by the material properties, process conditions and die geometry. The used methodology was commercial DEFORM-2D (two-dimensional). This finite element code was focus on the plastic deformation behavior of material with internal defective voids. These simulations were investigated, the damage factor distributions, stress-strain distributions around the defective voids, and the void dimensions. The result found, the effective stress distribution is clearly greater in the void region of the billet in the inner and outer of the channel. The mesh element distribution was significantly increasing in the 2-turn ECAE. Jun *et al* (2008) studied about the die structure optimization of equal channel angular extrusion for AZ31 magnesium alloy based on finite element method. In this analysis DEFORM-3D finite element code was used. The die was designed in three dimensional geometric models with different angles and with or without inner round fillets in the bottom. Important process parameters were calculated using above finite element code. The process parameters were stress-strain data of compression test, temperature of die and billet, and friction coefficient on deformation process were discussed. The result found, the equivalent strains are increased by comparing the 3D FEM results with theoretical calculations because of thermal and friction conditions. The lubrication condition is important to plastic deformation. The deformation homogeneity caused by fillets of the outer corner is larger. Eivani *et al* (2008) studied about the dead metal zone formation on strain and extrusion force during equal channel angular extrusion. The deformation of material during ECAE was analyzed using upper bound model. The ECAE considered the effect of die angle and friction between the samples and die walls on the geometry of dead metal zone formation. The result found, the friction coefficient between the material and die wall was more. When increasing the die angle, the critical value for friction coefficients increased that means formation of dead metal zone formation was less. The larger deformation zone develops with increasing the friction coefficient and decreasing the die angle. The total strain not only dependent on the die angle but also depend on friction coefficient.

Patil *et al* (2008) studied about the influence of friction in equal channel angular pressing. The friction between die and work piece has most influence on the on the extrusion pressure and flow in the process. This analysis was carried out using the ABAQUS software. The three dimensional finite element analysis was using for different values of coefficient of friction to understand the influence on material flow, pressure and strain homogeneity in equal channel pressing. The result found this analysis was, if the friction was increased the corner gap decreased due to back pressure. The inhomogeneity in strain distribution decreases with increase in friction until the backpressure is just sufficient to fill the corner gap. Patil *et al* (2010) studied about the influence of outer corner radius in equal channel angular pressing. In this study the plastic deformation and strain distribution was necessary for understanding the relationships between strain homogeneity and die geometry. In this study, the three dimensional simulation of ECAP process was carried out for six outer corner radii with channel angle of 105^0. The result found from above analysis was, the outer corner has a significant influence on the strain distribution in the body of work piece. The strain inhomogeneity was found to be high for both sharp and large outer corner dies. The average strain was decreased towards the head of the work-piece for all cases. Egrahimi (2010) studied about the investigation of strain behavior in the modified equal channel angular pressing die by 3D finite element code. ECAP process modification and die shape modification were carried out. The modification was the process to be continues for several passes, resulting a high homogeneity of strain distribution by using route C and the second modification was to obtain high magnitude of effective strain and high uniform strain distribution during ECAP process. In this analysis the simple and modified die was used to press the aluminum up to four passes. The result of the analysis was, the in-homogeneity of strain distribution was decreased in the modified ECAP die due to increasing the number of passes.

Li *et al* (2010) studied about the selection of outlet channel length and billet length in equal channel angular extrusion process. In this work, the finite element analysis was used. Deformation behavior of the material during the equal channel angular extrusion of a typical strain hardening with different combinations of outlet channel length and billet length were simulated. The result found, evaluated in terms of strain heterogeneity along the longitudinal direction, punch load-displacement curves, shape of the deformed billet. The shorter outlet channel leads to a longer steady state region and lower working load. The effect of outlet channel length was attributed to the friction forces in the outlet channel. The steady-state region in the billet increases with the billet length-to-width ratio until the ratio reaches a critical value. Hong-jun *et al* (2010) studied about the die structure optimization of equal channel angular extrusion for AZ31 magnesium alloy based on finite element method. In this work the three-dimensional (3D) geometric models, with different die angles ($90°$-$120°$) and with or without inner round fillets in the bottom die were designed. The ECAE process parameters were calculated in DEFORM-3D software, such as stress, strain data of compression test for AZ31 magnesium, temperatures of die billet and friction coefficient. The result found, the strain was decreased with increasing the fillets at outer corner. The deformation strain homogeneity was produced by fillets at outer corner increased compared with the die without fillets. Jiang *et al* (2012) studied about the equal channel angular extrusion of Ni Ti shape memory alloy tube. This tube was investigated by means of experiment, finite element code and microscopy test were conducted. In finite element analysis multiple coupled boundary conditions and multiple constitutive models were used. Based on experimental tests, the effective stress field, effective strain field and the velocity field were obtained. The result found, the effective strain field indicates the plastic deformation of the inner corner was greater than its outer corner. And the velocity field indicated the flow of material. The finite element analysis results were fine compared with experimental tests.

Jia-yong *et al* (2012) studied about the finite element analysis of die geometry and process conditions effects on equal channel angular extrusion for β-Titanium alloy. In this work, the finite element code DEFORM-3D was used. The equal channel angular extrusion process was conducted, to extrude the material without changing the geometrical shape. The result found, if the outer radius of the die geometry was increased, the magnitude of the equalent plastic strain was decreased. The inner radius of the die was increased; the equalent plastic strain deformation was reduced. The friction coefficient was increased with the plastic strain. If the ram speed was faster, the homogeneity of the equalent plastic strain was lower. The effect of temperature is higher; the plastic strain distribution was lower. Qiu *et al* (2012) studied about the plastic deformation of crystalline polymer materials in the equal channel angular extrusion process. It is a capable of transmit large amount plastic strain to bulk material through the application of uniform simple shear. Mainly material properties die geometry and process conditions were influence the shear deformation behavior during extrusion process. In this work finite element code abaqus was used. This finite element code was analyzed, the deformation behavior of the material, frictional conditions and strain speed of the process. The result of this work, the optimal strain homogeneity in the sample with lower dead zone formation, without involving any determinate effects can be achieved with channel angle and outer corner angle. Sharma *et al* (2014) studied about the multi pass equal channel angular pressing using 3D finite element analysis. It is a most important technique, used to produce an ultrafine grained structure in materials and to improve the mechanical and physical properties. In this work, four passes of the ECAP process was done. The billet was carried out at routes A, B_A, and C. Different channel angles and different values of coefficient of friction were used, to investigate their influence on the billet. The result of this work, the 90^0 channel angle was performed higher equalent plastic strain compared to other channel angles. The ECAP was a practical way for producing ultra fine grains in bulk metals.

2.4 Other Studies on Equal Channel Angular Extrusion Process

El-Sayed Sgerif *et al* (2013) studied about the effect of ECAP passes on the corrosion behavior of aluminum alloy (AA 1050) in natural sea water had investigated. In this work, to increasing the ECAP pass number up to 16 on the corrosion of AA 1050 in Arabian Gulf seawater. The work was carried out by using cyclic potentiodynamic polarization, chronoamperometric current-time variations, and electrochemical impedance spectroscopy. The AA 1050 after 20 min and 10 days immersed in the AGS solutions at room temperature. The result found, the annealed AA 1050 (0 passes) suffers both uniform and pitting corrosion after 20 min immersion in the AGS solution. An ECAP alloy showed lower corrosion, absolute current densities and higher polarization.

CHAPTER 3
METHODOLOGY

In this chapter we are going to discuss about the experimental and simulation work which consists of determination of material properties for aluminum, external die geometry, back pressure applications, friction, strain and strain homogeneity, and force during the process etc.

3.1 Experiment Setup

The equal channel angular extrusion tool consists of die, punch and billet. In ECAE process the die consisted of two channels of same cross section, the punch placed over the die. Before processing, the billet and walls of the die were coated with anti-seize lubrication in order to minimize the friction during ECAE process. The punch was lubricated which is used to press the material from the die. This was done by forcing the material through a die consisting of two channels of equal cross sectional area that intersect at an angle of Φ. There was also an additional angle Ψ which defines the arc of curvature at the outer point of the two intersecting channels.

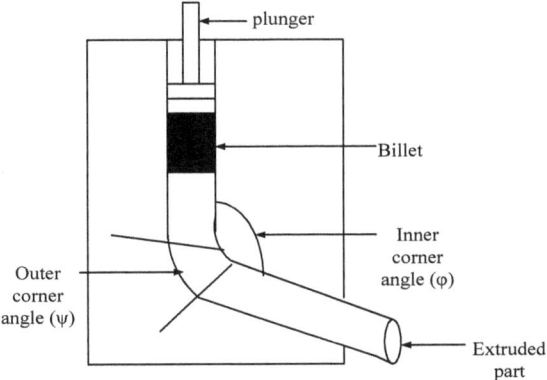

Figure 3.1: Schematic representation of ECAE setup

The dimensions of ECAE die for 9 mm circular work pieces were designed with φ (intersect angle) = 115°,125°and135° and ψ (the angle subtended by the arc of curvature at the point of intersection) = 6°. 3.1 shows the schematic of ECAE and photograph of the fabricated ECAE die fitted to the Universal testing machine of 200 tones capacity, respectively.

19

3.2 Equipment

The whole experimental investigation were done using "FIE Electronic Universal Testing machine (UTM)", model UTS-100 which can be used for conduction test in tension, compression and transverse test of metals and other material.

The UTM consists of three major parts:
- Machine frame or loading unit
- Hydraulic system
- Electronic control panel

3.2.1 Machine Frame or Loading Unit

Machine frame and loading unit consist of two cross heads and one lower table. Center cross head are adjustable by means of geared motor. Compression test is carried out between center and lower table while tension test is carried out between center table and upper cross head. Load is sensed by means of precision pressure transducer of strain gauge type. Loading unit is shown in Fig 3.2.

Figure 3.2: Universal Testing Machine fixtures

3.2.2 Hydraulic System Unit

Hydraulic system unit consists of motor pump unit with cylinder and piston. Safety valve is provided for additional safety as shown in Fig 3.3.

Figure 3.3: UTM Hydraulic system unit

3.2.3 Electronic Control Unit

Electronic control unit control the process by controlling the input parameter like load rate, strain rate, maximum load etc.

3.3 Die Design Details

The dimensions considered for developing the ECAE process setup in below table 3.2. The ECAE die contains two channels, equal in cross-section, intersecting at an angle near the center of the die. The test sample is machined to fit within these channels and it is pressed through the die using a punch. Die material was mild steel and the plunger was HSS steel (normalized condition), the normalizing of HSS steel was carried out by heating above the upper-critical-temperature line followed by cooling in still air to room temperature. The purpose of normalizing is to produce harder and stronger than annealing, so that for some applications normalizing may be a final heat treatment.

TABLE 3.1: TOOL DIMENSIONS AND PROCESSING PARAMETERS

Process Parameter	Value
Punch diameter (mm)	10 diameter, 75 height
Die channel angles	105°, 115°, 125°, 135°
Die outer Radii (mm)	6
Billet size (mm)	9 diameter, 70 height

3.4 Die Design Consideration

For die design of equal channel angular extrusion following factors are to be considered as given below.

- Desired shape of the product
- Material
- Billet size
- Process capacity
- Die material
- Extrusion pressure
- Extrusion temperature
- Heat treatment of punch material

3.5 Different Parts of the Setup

ECAE process setup required the die with different angle passages and punch along with raw materials. Die and punch are shown in the Fig.3.4. Die and Punch were manufactured with mild steel and hardened steel respectively. Dimensions are maintained as per the standards given in the literature.

Figure 3.4: Punch and different channel angles of die

3.6 Equal Channel Angular Extrusion Process Parameters

In equal channel angular extrusion process the strain homogeneity and the amount strain induced depends on the various parameters of the tool which are described below.

3.6.1 Channel Intersection Angle (Φ)

The Angle between the two intersecting channel refers to the channel intersection angle (Φ). The channel angle, Φ, is the most significant experimental factor since it indicate the total strain imposed in each pass and thus it has a direct influences on the nature of the as-pressed microstructure.

3.6.2 Outer Corner Angle (Ψ_0)

The angle of curvature, $\Psi 0$, denotes the outer arc where the two parts of the channel intersect within the die. This angle plays only a minor role in determining the strain imposed on the sample.

3.6.3 Inner Corner Angle (Ψi)

The angle of curvature, ψ_i, denotes the inner arc in the inner corner where the two parts of the channel intersect within the die. The inner corner angle (ICA) is one of the major factors affecting deformation homogeneity in work pieces during equal channel angular pressing. Inner corner angle (ICA) is one of the major factors affecting deformation homogeneity in work pieces during equal channel angular pressing.

3.6.4 Inner Corner Radius (R_i)

The radius, R_i given to the inner curvature in the inner corner is termed as inner corner radius.

3.6.5 Outer Corner Radius (R_o)

The radius, R_o given to the outer curvature in the outer corner is termed as outer corner radius. Outer corner has a significant influence on the strain distribution in the body of work-piece. Based on in homogeneity and average strain criteria, there is an optimum outer corner radius.

3.6.6 Temperature of Billet and Die

The pressing temperature is a key factor in any use of ECAP because it can be controlled relatively easily. It is generally experimentally easier to press specimens at high temperatures; optimum ultrafine-grained microstructures will be attained when the pressing is performed at the lowest possible temperature where the pressing operation can be reasonably conducted without the introduction of any significant cracking in the billets. By maintaining a low pressing temperature, this equilibrium grain size and the highest fraction of high-angle boundaries.

3.6.7 Friction

The friction between the die and the material has significant influence on the strain homogeneity and the force applied.

3.6.8 Force

Processing by ECAP is usually conducted using high-capacity hydraulic presses that operate with relatively high ram speeds. Typically, the pressing speeds are in the range of 1–20 mm/s. The pressing speed or force has significant influence on strain homogeneity.

3.6.9 Number of Passes

The number of passes represents the total strain induced in the material.

3.6.10 Deformation Route

When the sample is pressed through several consecutive passes, the shearing characteristics may be changed by rotating the sample between each pass. Thus the route with which the sample was re-entered to the ECAP die in each pass had an influence on the microstructure achieved due to successive change of the shear plane. There are four basic processing routes in ECAP.

Route A: 0°, all passes; the billet is not rotated between successive passes.
Route B Or BA: (90°), N even, 270° N odd; the billet is rotated 90° clockwise and counterclockwise alternatively.
Route C: (180°), all passes; the billet is rotated 180°.
Route D or BC: (90°), all passes; the billet is rotated 90° clockwise.

3.7 Modeling of Equal Channel Angular Extrusion Setup

Using Computer aided design (CAD) tool, CATIA, the equal channel angular extrusion setup was developed in three dimensions. The equal channel angular extrusion setup consists of a punch, billet and die. In modeling the die consists of three different die angles were considered. The die has made with inner corner angles of angles (φ) 115°, 125°, and 135° as shown in Fig. 3.4. The parts of the equal channel angular extrusion tool were designed in the part modeling module in CATIA. The punch, die modeled in CATIA were shown in the Figures 3.5, 3.6 and complete assembly is shown in the Fig. 3.7.

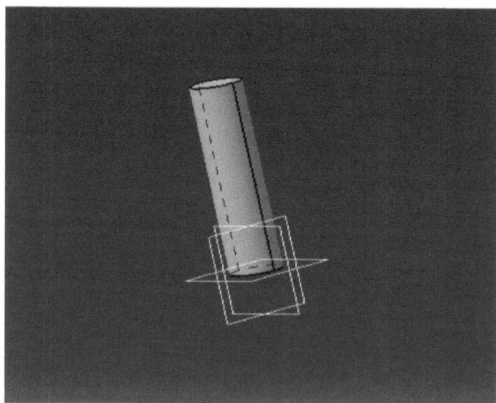

Figure 3.5: Modeling of Punch

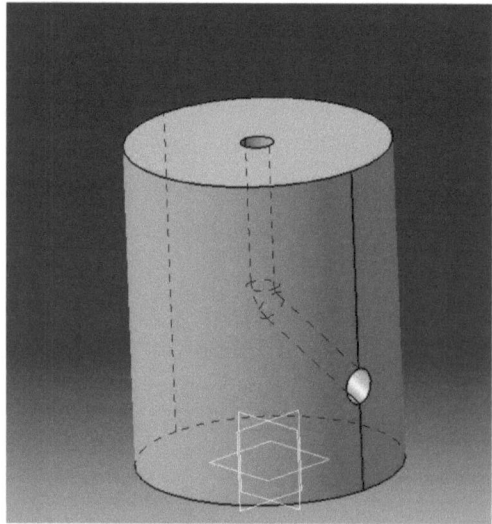

Figure 3.6: Modeling of Die

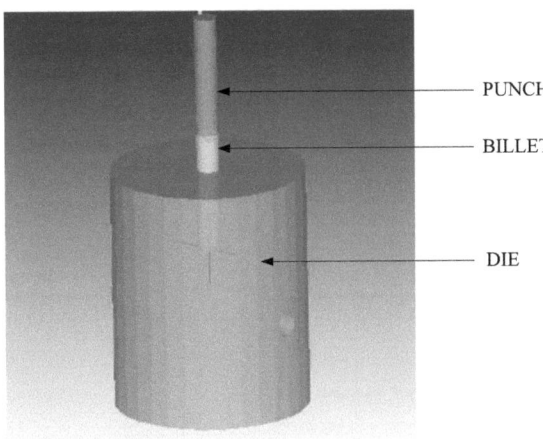

Figure 3.7: Assembly of ECAE setup

3.8 Raw Material Used For Experiment

Aluminum (Al) is the 3rd most abundant element in the earth's crust, ranking only behind oxygen and silicon. It makes up about 9% of the earth's crust, making it the most abundant of all metals. Accordingly, it is employed in a wide range of applications, mainly involving transportation, packaging, roofing, siding, door frames, screens, electrical appliances, automobile engines, heating and cooling systems, water purification, sewage treatment, etc.

The key advantages of using aluminum:
- Aluminum is light weight metal, non-magnetic and heat treatable.
- The specific gravity of Aluminum is one-third that of low carbon steel.
- The specific strength or the strength to weight ratio of Aluminum is excellent.
- Aluminum is rust free which makes it more useful in manufacturing rust resistant parts.
- It is used to create a corrosion resistant and a decorative surface finish by anodizing the metal.

TABLE 3.2: MATERIAL PROPERTIES OF THE ALUMINUM1100

Properties	
Density (x1000 kg/m^2)	2.71
Poisson's Ratio	0.33
Elastic Modulus (GPa)	70-80
Tensile Strength (MPa)	110
Yield Strength (MPa)	105
Elongation (%)	12
Hardness (HB500)	28
Shear Strength (MPa)	69

3.9 Experimental Procedure For ECAE

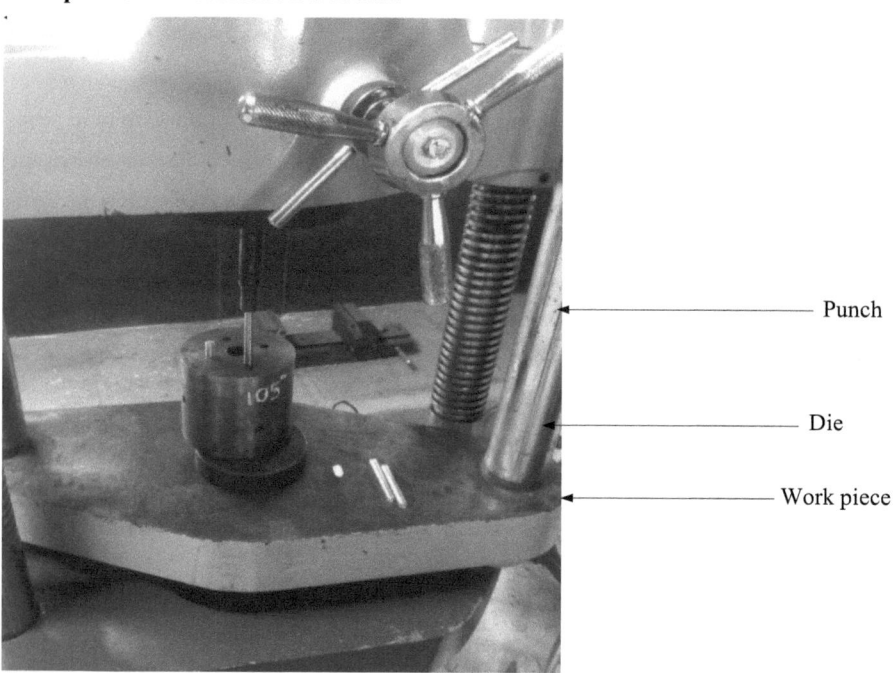

Figure 3.8: ECAE setup on Universal Testing Machine

Before starting the test the die, punch and inside face of extrusion chamber were cleaned. The full assembly of ECAE die was placed in between the base plate and centre table of Universal testing machine as shown in the Fig. 3.8. It was done so that extrude product would get enough clearance, when it comes out from the die. For carrying out an ECAE test the aluminum specimen was placed inside the die. Pure aluminum rods with 70mm×9mm (H×D) were taken. The punch was then inserted into its position. Machine was started and ECAE process was continued. Punch load was recorded at every 1mm movement of punch travel, which was read from computer fitted to the UTM. The application of the load was continued, the work piece till reaches out of the die. At this position machine was stopped and test was terminated. Experiments were conducted for three different die angles 105°, 115°, 125° and 135°.

3.10 Finite Element Analysis

In the present work finite element simulations were carried out using DEFORM–3D. Finite element analysis modeling is done using DEFORM-3D Version. DEFORM-3D is a Finite Element Method (FEM) based process simulation system designed to analyze various forming and heat treatment processes. By simulating manufacturing processes on a computer, this advanced tool allows designers and engineers to:

- Reduce the need for costly shop floor trials and redesign of tooling and processes
- Improve tool and die design to reduce production and material costs
- Shorten lead time in bringing a new product to market

Unlike general purpose FEM codes, DEFORM is tailored for deformation modeling. A user friendly graphical user interface provides easy data preparation and analysis so engineers can focus on forming.

DEFORM–3D processes has the three major components such as

(i) Pre-processor: It creates the part geometry, assembles the part geometries, generates mesh, inputs the material data and processes the data required to analyze the simulation for generating the required database file.

(ii) Simulation engine: It performs the numerical calculations required to analyze the process and writing the results to the database file. It performs the actual simulation calculations. It is the main processor.

(iii) Post processor: It reads the database file from the simulation engine and displays the results graphically, extracts the numerical data and derivative required quantities.

DEFORM provides different methods of defining the flow stress. For the present simulations, the power law was used which can be defined as:

$$\sigma = K\varepsilon^n$$

Where σ the flow stress of the material, K is the material constant, ε is the effective plastic strain and n is the strain exponent.

3.11 Simulation Methodology

FEM simulation was carried out using DEFORM-3D Version. As discussed above DEFORM-3D consist of three main components. In the pre-processor, the object of 9mm diameter and 70mm length defined. Generated the mesh with 20, 0000 element. The geometry of the ECAE die was imported in the form of STL file generated in CATIA. A material data was taken as per the literature. After setting the whole setup a database (DB) is generated.

The simulation engine reads the database file generated in pre-processor and performs the actual solution calculations, and writes the result in database file.

In post-processor, we evaluated all the results, and drawn the graphs, the strain hardening law and the plasticity model considered were the Hollomon's law. From numerous investigations, it was well-established that the stress-strain analogy can be used to determine the formability. Hollomon equation was one such equation that describes the strain-hardening behavior of metal deformation. The Hollomon equation was given by

$$\sigma = K\varepsilon^n \qquad (3.1)$$

Where σ is stress, K is the material strength coefficient, ε is the strain, and n is the material work-hardening exponent. As this equation does not consider strain rate, this was the simplest and most commonly used strain hardening equation. For maximum strength and strain hardening the materials that obey the Hollomon equation with large K value and n values between 0.1 and 0.3 are required However, various studies have established that the strain hardening behavior was strain-path, stress-state, strain-rate and temperature dependent.

CHAPTER 4
RESULTS AND DISCUSSION

In this chapter, all the results of Experimental investigation, FEM analysis are discussed. Results with different experimental conditions are compared. The numerical results are also compared with the experimental values. Effect of different die angles with respect to like load and displacement are also discussed. The punch force and displacement during the ECAE process for the Aluminum alloy are effectively evaluated from the results obtained from experiments and simulations.

4.1 Experimental Investigation
Figure 4.1 indicates the specimens used for the ECAE process.

 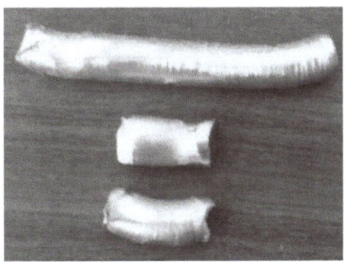

Before processing   After processing

Figure 4.1: Before and after ECAP processing of the Al samples

4.2 Stress Evaluation at Different Steps
The stress was evaluated at different steps of the work piece in numerical simulation at different die channel angles are shown form Fig. 4.2 to Fig. 4.5.

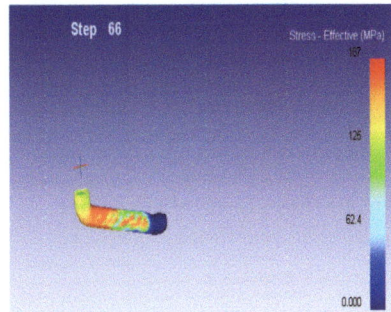

(a)

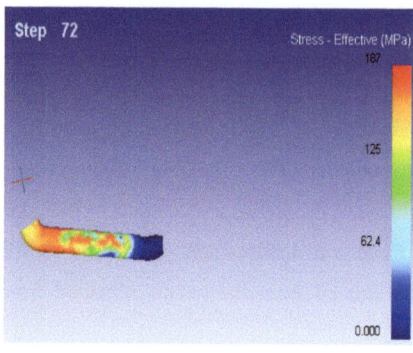

(b)

Figure 4.2: Stress formations at different progressions for die channel angle of 105°

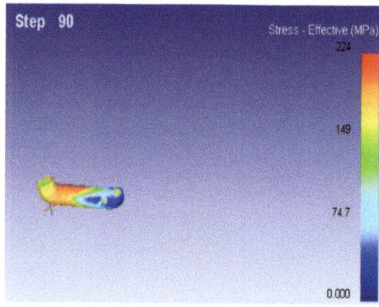

(a)

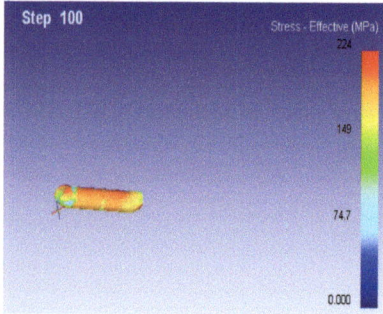

(b)

Figure 4.3: Stress formations at different progressions for die channel angle of 115°

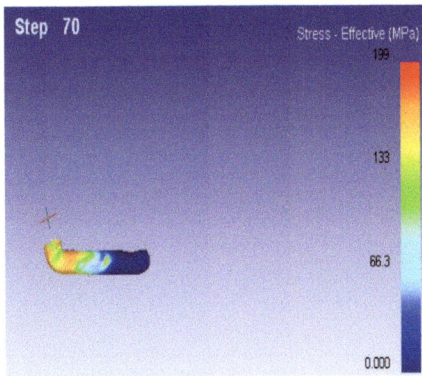

(a)

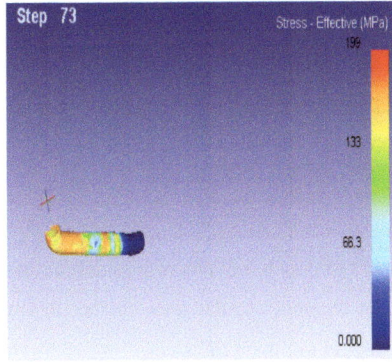

(b)

Figure 4.4: Stress formations at different progressions for die channel angle of 125°

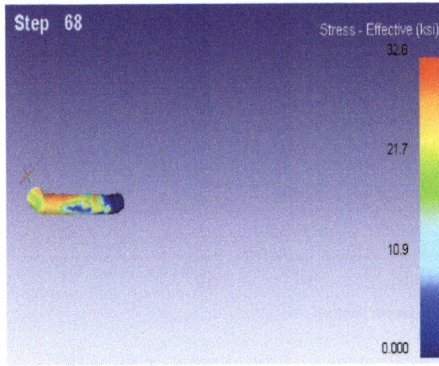

(a)

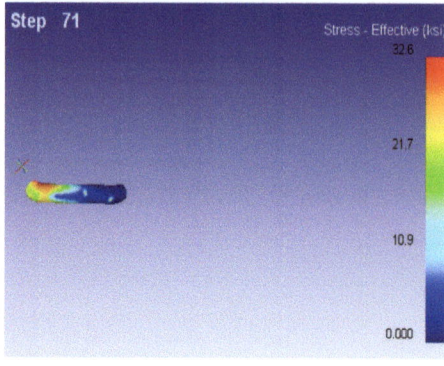

(b)

Figure 4.5: Stress formations at different progressions for die channel angle of 135°

From the above figures the maximum stress occurred at channel intersecting place.
- ➤ The maximum value of stress formed is 32.6
- ➤ The minimum value of stress formed is 0.00

4.3 Punch Force Evaluation

The variation of compressive extrusion load with respect to punch movement was determined from the ECAE test of round billet using different die channel angular dies. In the present investigation it is found that, the extrusion load with respect to punch travel.

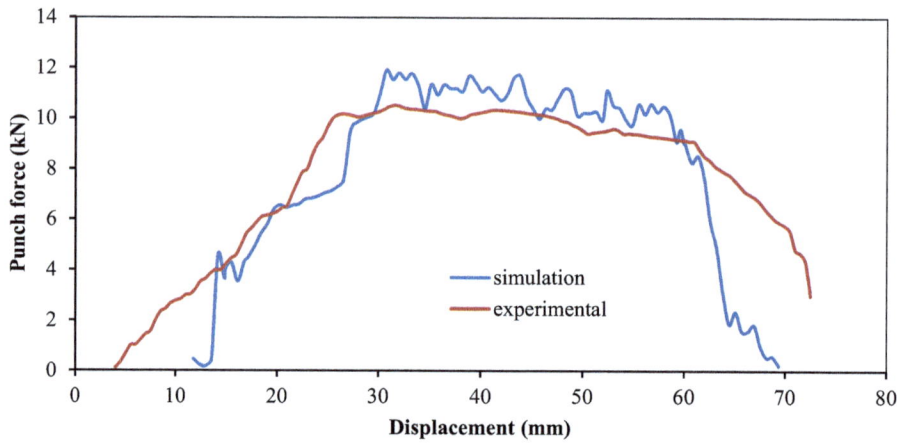

Figure 4.6: Comparision of extrusion load with displacement with die channel angle of 105°

Figure 4.6 is showing the punch force variation of both simulation and experimental data and from the graph we can observe that the punch force required in simulation is slightly higher when compared to the experiment. The maximum punch force in experiment was 11.78 kN and the simulation was 10.34 kN. The simulation results were good agreement with experimental results.

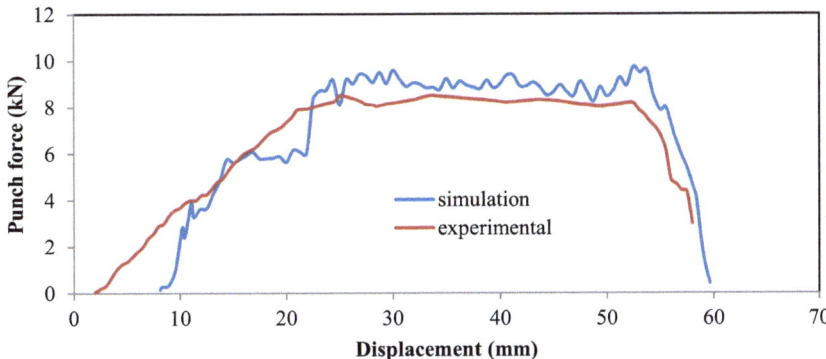

Figure 4.7: Comparision of extrusion load with displacement with die channel angle of 115°

Figure 4.7 shows the punch force variation of both simulation and experimental data and from the graph we can observe that the punch force required in simulation is slightly higher when compared to the experiment. The maximum punch force in experiment was 8.31kN and the simulation was 9.73 kN. The simulation results were good agreement with experimental results.

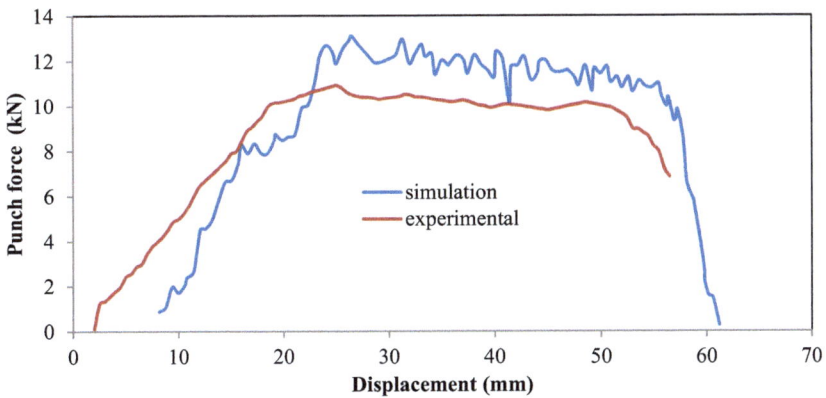

Figure 4.8: Comparision of extrusion load with displacement with die channel angle of 125°

35

Figure 4.8 indicated the punch force variation of both simulation and experimental data and from the graph we can observe that the punch force required in simulation is slightly higher when compared to the experiment. The maximum punch force in experiment was 10.43 kN and the simulation was 12.97 kN. The simulation results were good agreement with experimental results.

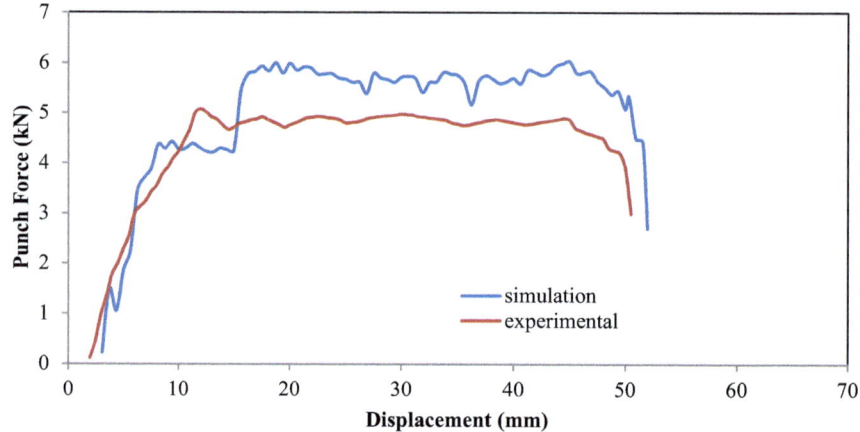

Figure 4.9: Comparision of extrusion load with displacement with die channel angle of 135°

The punch force variation of both simulation and experimental data can observe from Fig. 4.9 that the punch force required in simulation is slightly higher when compared to the experiment. The maximum punch force in experiment was 5.8 kN and the simulation was 6.03 kN. The simulation results were good agreement with experimental results.

From the above figures as the punch pushes the work piece through the die, the load applied increases linearly up till the point where in the work piece starts moving through the angled extrusion plane. At this point the temperature of the work piece rises thus causing a change in plasticity of the material due to thermal effects. But as the larger area of the work piece is still in contact with the cold surface of the die this thermodynamic deformation is limited. But gradually as the work piece is pushed further in thermal gradient of the work piece is more predominant thus effecting the material characteristics by transforming the solidity of work piece. At this stage the work piece plasticity is no longer linear and the difference in temperature gradient causes a slip among the grain surfaces causing a twist in the material as the work piece is completely pushed out. The above figures present the comparison of force requirement for the above mentioned configurations. Numerical results are a little higher compared to the experimental observations. The nature of the load-displacement curve after the peak in experimental study is slightly different (i.e. the load does not remain steady, but reduces with displacement). It is believed that this may be due to the assumption in the numerical analysis that the friction coefficient remains constant, which is not the real case. In reality due to continuous change in frictional contact area, and heating during sliding motion, the dynamic coefficient of friction must vary. In addition, the force measurement is not perfect. Using a simple assembly, a fish scale measures one fourth of the actual load, and thus any small error in reading the gauge is amplified.

TABLE 3: DIE CHANNEL ANGLES AND THEIR OBTAINED RESULTS

S. No	Die channel angle	Max. Stress	Min. Stress	Experimental Punch Force	Simulation Punch Force
1	105°	187	62.4	11.78	10.34
2	115°	224	74.7	8.31	9.73
3	125°	199	66.3	10.43	12.97
4	135°	32.6	10.9	5.8	6.03

CHAPTER 5
CONCLUSIONS AND FUTURE SCOPE

The current work aimed to contribute the literature of ECAE process and their processing technologies. The parameters used in simulation and experimental and their effects on the results are investigated. The effects of punch force and stresses are evaluated in different die channel angles during the ECAE process on Al alloy. The following conclusions are drawn from the results.

- The experimental punch force is more in 105° die channel angle.
- The experimental punch force is less in 135° die channel angle.
- The simulation punch force is more in 125° die channel angle.
- The simulation punch force is less than 135° die channel angle.
- The maximum stress occurred in 115° die channel angle.
- The minimum stress occurred in 135° die channel angle.

Future Scope

A possible future study about the ECAE process of aluminum alloy may include the simulation and experimental of multi pass shear deformation process may be carried out to investigate the influence of the process variables in detail and to better identify optimum conditions for better results. Microstructural evolution, evolution of grain subdivision and cell formation mechanisms hardness also investigate in future study.

REFERENCES

1. **V.M. Segal** (1999) Slip line solutions, deformation mode and loading history during equal channel angular extrusion *Materials Science and Engineering* A345 36_ 46.
2. **V.M. Segal** (2003) Equal channel angular extrusion: from macro mechanics to structure formation *Materials Science and Engineering* A271 322–333.
3. **Q. Wei, T. Jiao, S.N. Mathaudhu, E. Ma, K.T. Hartwig, K.T. Ramesh** (2003) Microstructure and mechanical properties of tantalum after equal channel angular extrusion (ECAE) *Materials Science and Engineering* 266-272.
4. **M.V. Markushev, M.Yu. Murashkin** (2004) Structure and mechanical properties of commercial Al–Mg 1560 alloy after equal-channel angular extrusion and annealing *Materials Science and Engineering* A 367 234–242.
5. **Y.Y. Wang, P.L. Sun, P.W. Kao, C.P. Chang** (2004) Effect of deformation temperature on the microstructure developed in commercial purity aluminum processed by equal channel angular extrusion *Scripta Materialia* 613–617.
6. **M Saravanan, R M Pillai, B C Pai, M Brahmakumar and K R Ravi** (2006) Equal channel angular pressing of pure aluminum, bulliten master science 679-689.
7. **D.R. Fang, Z.F. Zhang, S.D. Wu, C.X. Huang, H. Zhang, N.Q. Zhao, J.J. Li** (2006) Effect of equal channel angular pressing on tensile properties and fracture modes of casting Al–Cu alloys, *Materials Science and Engineering* 305–313.
8. **A.A. Gazder, S. Li, F.H. Dalla Torre, I.J. Beyerlein, C.F. Gu, C.H.J. Davies, E.V. Pereloma** (2006) Progressive texture evolution during equal channel angular extrusion *Materials Science and Engineering* 259–267.
9. **A.V. Nagasekhar, Uday Chakkingal and Venugopal** (2006) Equal channel angular extrusion of tubular aluminum alloy specimens-analysis of extrusion pressures and mechanical properties *Journal of manufacturing processes* 112-120.
10. **W. Yiu Yeung, R. Wuhrer, M. Cortie and M. Ferry** (2007) Equal channel angular extrusion of high purity gold *Materials forum* 31-35.
11. **S.Suzuki, H.Utsunomiya, H.Nakajima** (2008) Equal-channel angular extrusion process of lotus-type porous copper *Materials Science and Engineering* A490 465–470.
12. **M.H. Paydar, M. Reihanian, E. Bagherpour, M. Sharifzadeh, M. Zarinejad, and T.A. Dean** (2008) Consolidation of Al particles through forward extrusion-equal channel angular pressing (FE-ECAP) *Materials Letters* 3266–3268.

13. **L.L. Gao, X.H. Cheng** (2008) Microstructure and dry sliding wear behavior of Cu–10%Al–4%Fe alloy produced by equal channel angular extrusion *Wear* 265 986–991.
14. **R.Daly, S.Zghal, N.Njeh** (2009) Effects of annealing on the microstructure and properties of cua1copper processed by Equal Channel Angular Extrusion *Physics Procedia* 677–684.
15. **Atef Rebhi, Thabet Makhlouf, Nabil Njah, Yannick Champion, Jean-philippe Couzinie** (2009) Characterization of aluminum processed by equal channel angular extrusion: effect of processing route *Materials characterization* 1489–1495.
16. **Nguyen Q. Chinh, Jeno Gubicza, Tomasz Czeppe, János Lendvai and Terence G. Langdon** (2009) Processing age-hardenable alloys by equal-channel angular pressing at room temperature: strategies and advantages *Materials Science Forum* Vols. 633-634 527-534.
17. **Somjeet Biswas, Satyaveer Singh Dhinwal, Satyam Suwas** (2010) Room-temperature equal channel angular extrusion of pure magnesium Acta Materialia 58 3247–3261.
18. **Vasile Danuţ, Cojocaru, Doina Raducanu, Nicolae Serban, Ion Cinca, Rami Saban** (2010) Mechanical behavior comparison between unprocessed and ECAP (equal channel angular pressing) processed 6063-t835 aluminum alloy *Bullitin series* 72 193-202.
19. **Dong-Hwan Kang, Tae-Won Kim** (2010) Mechanical behavior and micro structural evolution of commercially pure titanium in enhanced multi-pass equal channel angular pressing and cold extrusion *Materials and Design* 31 S54–S60.
20. **Dmitry Orlov, George Raab, Torbjorn T. Lamark, Mikhail Popov, Yuri Estrin** (2011) Improvement of mechanical properties of magnesium alloy ZK60 by integrated extrusion and equal channel angular pressing *Acta Materialia* 59 375–385.
21. **P. Eslami, A. Karimi Taheri** (2011) An investigation on diffusion bonding of aluminum to copper using equal channel angular extrusion process *Materials Letters* 65 1862–1864.
22. **Kazeem O. Sanusi, Oluwole D. Makinde Graeme J. Oliver** (2012) Equal channel angular pressing technique for the formation of ultra-fine grained structures *Research article* 1-7.
23. **Mohan Reddy, Santhosh Kumar M, Venkata Ajay Kumar G** (2013) Improving Mechanical Properties of aluminum 7075 alloy by Equal Channel Angular Extrusion process International Journal of Modern Engineering Research 2713-2716.
24. **Z.Q. Fan, T. Hao, S.X. Zhao, G.N. Luo, C.S. Liu, Q.F. Fang** (2013) The microstructure and mechanical properties of T91 steel processed by ECAP at room temperature Journal of Nuclear Materials 434 417–421.

25. **Sai-yi LI** (2013) Application of crystal plasticity modeling in equal channel angular extrusion *Transaction of non ferrous metals society of China* 23. 170-179.
26. **T. Hao, Z.Q. Fan, S.X. Zhao, G.N. Luo, C.S. Liu, Q.F. Fang** (2013) Microstructures and properties of ultrafine-grained tungsten produced by equal-channel angular pressing at low temperatures *Journal of Nuclear Materials* 433 351–356.
27. **Hao Li, Saiyi Li, Donghong Zhang** (2010) On the selection of outlet channel length and billet length in equal channel angular extrusion *Computational Materials Science* 49 293–298.
28. **F. Bouaksa, C. Ovalle Rodas, F. Zaïri, G. Stoclet, M. Naït-Abdelaziz, J.M. Gloaguen, T. Tamine, J.M. Lefebvre** (2014) Molecular chain orientation in polycarbonate during equal channel angular extrusion: Experiments and simulations *Computational Materials Science* 85 244–252.
29. **Y.X. Tong, P.C. Jiang, F. Chen, B. Tian, L. Li, Y.F. Zheng, Dmitry V. Gunderov, Ruslan Z. Valiev** (2014) Microstructure and Martensitic transformation of an ultrafine-grained TiNiNb shape memory alloy processed by equal channel angular pressing *Intermetallics* 49 81-86.
30. **Raja Thiyagarajan1, A.Gopinath** (2014) Enhancement of Mechanical Properties of AA 6351 Using Equal Channel Angular Extrusion (ECAE) *Materials Science and Metallurgy Engineering* Vol. 2, No. 2, 26-30.
31. **Irene Beyerlein, Carlos N.Tome** (2004) analytical modeling of material flows in equal channel angular extrusion process *Materials Science and Engineering* 171–190.
32. **B. Aour, F. Zaıri, J.M. Gloaguen, M. Naıt-Abdelaziz, J.M. Lefebvre** (2006) Numerical investigation on equal channel angular extrusion process of polymers, Computational Materials Science 491–506.
33. **A.R.Eivani, A.Karimi Taheri** (2008) the effect of dead metal zone formation on strain and extrusion force during equal channel angular extrusion, *Computational Materials Science* 14–20.
34. **A.R. Eivani, A. Karimi Taheri** (2008) Effective strain based on shear and principal strains in equal channel angular extrusion with outer curved corner, *Computational Materials Science* 41. 409- 419.
35. **T. Bajor, M. Krakowiak, P. Szota** (2014) numerical analysis of AZ61 magnesium alloy extrusion process by modified equal channel angular extrusion (ECAE) method Metalurgija 53 4, 485-488 485.

36. **Hyoung Seop Kim , Min Hong Seo, Sun Ig Hong** (2000) On the die corner gap formation in equal channel angular pressing *Materials Science and Engineering* A291 (2000) 86–90.
37. **S. Li, M.A.M. Bourke, I.J. Beyerlein, D.J. Alexander, B.Clausen** (2004) Finite element analysis of the plastic deformation zone and working load in equal channel angular extrusion, *Materials Science and Engineering* 217–236.
38. **A.V. Nagasekhar, Yip Tick-Hon** (2004) optimal tool angles for equal channel angular extrusion of strain hardening materials by finite element analysis *Computational Materials Science* 489–495.
39. **A.V. Nagasekhar, Yip Tick-Hon, S. Li, H.P. Seow** (2006) Stress and strain histories in equal channel angular extrusion/pressing *Materials Science and Engineering* 143–147.
40. **R. Luri, C. J. Luis, M. A. Sebastian** (2006) A new configuration for equal channel angular extrusion dies, *Manufacturing engineering* 860-865.
41. **F. Zairi, B. Aour, J.M. Gloaguen, M. Naït-Abdelaziz , J.M. Lefebvre** (2006) Numerical modeling of elastic-viscoplastic equal channel angular extrusion process of polymer, *Computational Materials Science* 202–216.
42. **F. Zaırı, B. Aour, J.M. Gloaguen, M. Naıt-Abdelaziz and J.M. Lefebvre** (2007) Influence of the initial yield strain magnitude on the materials flow in equal channel angular extrusion process *Scripta Materialia* 105–108.
43. **A.V.Nagasekhar, Yip Tick-Hon, H.P. Seow** (2007) Deformation behavior and strain homogeneity in equal channel angular extrusion/pressing *journal of materials Processing Technology* 449–452.
44. **Dyi-Cheng Chen, Ching-Pin Chen** (2008) Investigation into equal channel angular extrusion process of billet with internal defects, *journal of materials processing technology* 419–424.
45. **HU Hong-jun, ZHANG Ding-fei, PAN Fu-sheng** (2008) Die structure optimization of equal channel angular extrusion for AZ31 magnesium alloy based on finite element method (2010) *Transaction of nonferrous metals society of china,* 20.259-266.
46. **Basavaraj V Patil, Uday Chakkingal, T S Prasanna Kumar** (2008) Influence of friction in equal channel angular pressing – a study with simulation, *Metal* 13-15.

47. **Basavaraj V. Patil, Uday Chakkingal and T. S. Prasanna Kumar** (2010) Influence of outer corner radius in equal channel angular pressing *World Academy of Science, Engineering and Technology* 898-904.
48. **Faramarz Djavanroodi and Mahmood Ebrahimi** (2010) Investigation of strain behavior in the modified equal channel angular pressing die by 3D finite element method *Journal of applied sciences* 2411-2418.
49. **Shu-yong Jiang, Ya-nan Zhao, Yan-qiu Zhang, Ming Tang, Chun-feng li** (2013) Equal channel angular extrusion of NiTi shape memory alloy tube *Transaction of Nonferrous Metals Society China* 23 2021−2028.
50. **SI Jia-yong, GAO Fan, ZHANG Ji** (2012) Finite Element Analysis of Die geometry and Process Conditions effects on equal channel angular extrusion for β-Titanium Alloy *Journal of iron and steel research international*, 19(10) I 54-58.
51. **Jianhui Qiua, Takuya Murataa, Xueli Wua, Masayoshi Kitagawa, Makoto Kudo** (2012) Plastic deformation mechanism of crystalline polymer materials in the equal channel angular extrusion process, *Journal of Materials Processing Technology* 212,1528– 1536.
52. **Sanjeev Sharma, Ashok Kumar Raghav, Surendra Kumar** (2014) Study of multi pass equal channel angular pressing using 3D finite element analysis *International Journal of Emerging Research in Management &Technology* vol-3, 112-116.
53. **El-Sayed M. Sherif, Mahmoud S. Soliman, Ehab A. El-Danaf, A. A. Almajid** (2013) Effect of Equal-Channel Angular Pressing Passes on the Corrosion Behavior of 1050 Aluminum Alloy in Natural Seawater *International journal of electrochemical science* 8 1103 – 1116.

THESES

1. **Pinar Karpuz** (2005) Investigation of the effects of equal channel angular extrusion on light weight alloys *Middle East Technical University*.
2. **Rahul Rajendra Murudkar (2009)** Development of a continuous equal channel angular extrusion (ECAE) process *Texas A&M University*.